FLORA OF TROPICAL EAST AFRICA

GOODENIACEAE

F. G. Davies

Trees, shrubs or herbs, without latex. Leaves alternate, rarely opposite, or radical, exstipulate, simple, often obovate to spathulate. Inflorescence cymose or reduced to solitary axillary flowers. Flowers ⚥, protandrous, irregular. Calyx adnate to the ovary basally, distally free as a 5-toothed or entire limb. Corolla with a long tube split for part or all of its length and a limb divided into 5 subequal lobes. Stamens 5, from the base of the corolla, free or occasionally joined at the anthers; anthers dehiscing introrsely. Ovary superior to inferior, 1–2-locular, each containing 1 or several ovules; style emerging laterally from the split corolla-tube, bearing a cup-shaped ciliate indusium around the simple or bifid stigma. Fruit a capsule or drupe. Seeds with endosperm.

A family of 14 genera and over 300 species, mostly in Australia.

Cross-pollination is apparently favoured by the stamens releasing pollen into the cup of the indusium at the bud stage while the stigma becomes receptive much later.

SCAEVOLA

L., Mant. Pl. 2 : 145 (1771), *nom. conserv.*

Leaves alternate, rarely opposite, sessile or petiolate, entire or the margin toothed, glabrous or with stellate or simple hairs. Flowers solitary or in axillary bracteate cymes, variously coloured. Calyx-tube turbinate to globose; limb nearly entire to deeply 5-toothed. Corolla-tube split down one side to the base, often pubescent at least within; lobes subequal, spreading obliquely. Stamens free. Ovary inferior; style simple or 2-lobed. Fruit a drupe, the outer layer often succulent, the inner layer (endocarp) woody and enveloping the seeds.

About 80 species, 2 of which are widespread tropical strand plants.

Calyx-limb distinctly toothed, 3–6 mm. long (fig. 1/3);
  fruits whitish; leaf-axils with dense silky hairs (fig.
  1/2) . . . . 1. *S. sericea*
Calyx-limb scarcely split, 2 mm. long (fig. 1/11); fruits
  blue-black; leaf-axils glabrous or with sparse hairs
  (fig. 1/10) . . . . 2. *S. plumieri*

1. **S. sericea** *Vahl,* Symb. Bot. 2 : 37 (1791). Type : Cook Is., Niue [Savage I.], *G. Forster* (C, holo.)

Shrub or small tree, 2–3(–4) m. tall. Stem glabrous, fleshy, bearing persistent leaf-scars. Leaf-axils (and bract-axils) with prominent tufts of white silky hairs; leaves sessile, spathulate or obovate to narrowly obovate, 6–25 cm. long, 3–10 cm. broad, entire or distally faintly crenate, obtuse, narrowed gradually below the middle, abruptly broader at the base which

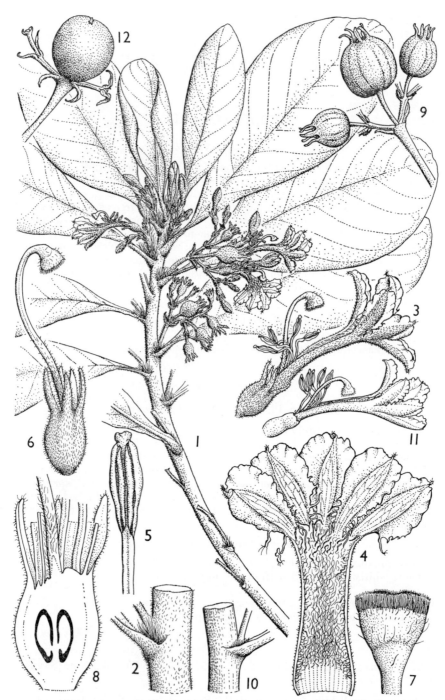

FIG. 1. *SCAEVOLA SERICEA*—**1**, flowering branch, × ⅔; **2**, leaf-axil, × 2; **3**, flower, × 2; **4**, corolla spread out, × 3; **5**, stamen, × 8; **6**, calyx and pistil, × 3; **7**, tip of style with indusium, × 6; **8**, longitudinal section of ovary, × 6; **9**, fruits, × 1. *S. PLUMIERI*—**10**, leaf-axil, × 2; **11**, flower, × 2; **12**, fruit, × 1. 1–9, from *Faulkner* 2681; 10–12, from *Faulkner* 2289. Drawn by Mrs M. E. Church.

partially clasps the stem, glabrous, thinly fleshy, light green with a tendency to turn blue when dried; lateral veins faint, all widely diverging, up to 25 pairs. Inflorescences axillary, cymose, 5–10 cm. long, 1–13-flowered, over-topped by the leaves, with many paired pubescent bracts 2–3 mm. long. Calyx-tube turbinate; teeth 3–6 mm. long, erect, persisting and sometimes lengthening on the fruit. Corolla blue to white, sometimes with mauve markings; tube 8–17 mm. long, 3 mm. broad; lobes elliptical, 8–13 mm. long, 4–5 mm. broad, the central portion thicker than the crenulate marginal zone. Ovary 2-locular; style simple. Fruit whitish, subglobose, 8–15 mm. long, slightly constricted medially, faintly 10-ribbed, 2-seeded. The corky fruits float in sea water and remain viable after prolonged immersion. Fig. 1/1–9.

KENYA. Kwale District: Diani Beach, 28 Aug. 1959, *Napper* 1298!; Mombasa I., English Point, 29 May 1934, *Napier* 3288! & Cliff Slopes, 28 Jan. 1953, *Drummond & Hemsley* 1059!

TANZANIA. Tanga District: Tangata, Mwarumau wa Ngamao, 10 Oct. 1943, *Lindeman* 1071!; Rufiji District: Mafia I., Adani, 9 Aug. 1937, *Greenway* 5040!; Zanzibar I., Pwani Mchangani, 26 Jan. 1929, *Greenway* 1203!; Pemba I., Tundaua, July 1901, *Lyne* 117!

DISTR. **K**7; **T**3, 6; **Z**; **P**; South Africa (Natal), Mascarene Is., India, China, Australia, Pacific Is.

HAB. Strand vegetation immediately above high water level, on sand and coral rock

SYN. [*Lobelia frutescens* sensu Mill., Gard. Dict., ed. 8 (1768), pro parte, *non* Mill. sensu stricto]
   L. taccada Gaertn., Fruct. 1: 119, t. 5 (1788). Type: seed formerly at Leiden, not found
   Scaevola koenigii Vahl, Symb. Bot. 3: 36 (1794); U.O.P.Z.: 443 (1949). Type: India, König (BM, isotype!)
   S. taccada Roxb., Hort. Beng.: 15 (1814). Type: plate 59 in Rheede, Hortus Malabaricus
   S. taccada (Gaertn.) Roxb., Fl. Indica 2: 146 (1824); K.T.S.: 230 (1961); Guillaumet in Fl. Mascareignes 110: 1, fig. 1/1–6 (1976), *non S. taccada* Roxb., *nom. illegit.*
   [*S. frutescens* sensu Krause in E.P. IV. 227: 125 (1912); T.T.C.L.: 237 (1949), *non* (Mill.) Krause sensu stricto]

VARIATION. The South African specimens have larger flowers and distinctly crenate leaves; they may constitute a subspecies.

2. **S. plumieri** (*L.*) *Vahl*, Sym. Bot. 2: 36 (1791); K.T.S.: 230 (1961); F.W.T.A., ed. 2, 2: 315 (1963); Guillaumet in Fl. Mascareignes 110: 2, fig. 1/7–9 (1976). Lectotype: plate 79 in Catesby, Nat. Hist. 1 (1731), of a plant from the Bahamas

A much-branched evergreen shrub or small tree, glabrous and succulent, with the leaves clustered at the branch-tips and leaving leaf-scars below. Stem yellowish greenish. Leaf-axils without or with sparse silky hairs; leaves sessile or petiolate (the petiole winged), obovate, 5–11 cm. long, 2–7 cm. broad, entire, obtuse, narrowed strongly from the widest part towards the base, thickly fleshy, yellowish green; lateral veins obscure, 2–4 pairs, the lower long and narrowly diverging from the midrib. Inflorescence a series of axillary cymes bearing conspicuous glabrous bracts (which may have hairs in their axils) and 1–7 sessile flowers. Calyx-limb short, scarcely divided, 2 mm. long. Corolla white or greenish; tube 10–12 mm. long, 3 mm. broad, crowded with hairs inside and shortly pubescent outside; lobes 6–10 mm. long, with a greenish, thick central part, and white, thin, crenulate marginal part. Ovary 2-locular, 1 ovule only developing. Fruit blue or black, sub-globose, 10–15 mm. across, very fleshy, drying warty. Seeds single, not dispersed from the fruit. Fig. 1/10–12.

KENYA. Kwale District: Twiga, 13 Jan. 1964, *Verdcourt* 3910!; Mombasa I., English Point, 29 May 1934, *Napier* 3287!; Kilifi District: Shanzu Beach, 26 Dec. 1967, *Greenway* 13134!

TANZANIA.   Uzaramo District: Kunduchi, 1 June 1966, *Welch* 623! & Ras Rongoni [Rangani] Point, 26 Nov. 1966, *Gillett* 18031!; Rufiji District: Mafia I., Ngombeni, 14 Aug. 1932, *Schlieben* 2669!; Zanzibar I., Chukwani, 18 June 1959, *Faulkner* 2289!

DISTR.   **K**7; **T**3, 6, 8; **Z**; coast of eastern Africa from Somalia to Cape Province (Algoa Bay) and of western Africa from S. Tomé to Angola, also Mascarene Is., Ceylon, and tropical America south from Florida

HAB.   Coastal sand-dunes

SYN.   *Lobelia plumieri* L., Sp. Pl.: 929 (1753), pro parte

     L. frutescens Mill., Gard. Dict., ed. 8 (1768), *nom. superfl.* based on *L. plumieri* L.

     Scaevola lobelia Murr. in Syst. Veg., ed. 13: 178 (1774); Hiern in F.T.A. 3: 462 (1877), *nom. superfl.* based on *Lobelia plumieri* L.

     S. thunbergii Eckl. & Zeyh., Enum. Pl. Afr. Austr.: 387 (1837).   Type: South Africa, Cape Province, Algoa Bay, *Ecklon & Zeyher* (K, iso.!)

NOTE.   In Réunion it has become an inland coloniser of lava flows, in pioneer vegetation.

INDEX TO GOODENIACEAE